Climate change

Dr. Hemant Pathak

Copyright © 2015 Dr. Hemant Pathak

All rights reserved.

ISBN: 1503363384
ISBN-13: 978-1503363380

DEDICATION

Dedicated to Shri Sainath Maharaj the all omnipotent of world the most merciful.

CONTENTS

	Foreword	5
	Glossary	7
1	Introduction	12
2	Human Causes	13
3	Natural causes	14
4	Climate forcing mechanisms	15
5	Effects of Climate Change	16
6	Climate models	21
7	Greenhouse gas reductions and alternate energy and Improved technology	21
8	Climate benefits	23
9	Conclusion	23
10	References	24

Foreword

The Earth's climate is changing. Scientists are highly confident that many of these observed changes can be linked to the climbing levels of carbon dioxide and other greenhouse gases in our atmosphere, which are caused by human activities. how much emissions are curtailed, the future could bring a relatively mild change in climate or it could deliver extreme changes that could last thousands of years.

Climate Change; provides a unique insight into the problems our planet faces in terms of temperatures are rising, snow and rainfall patterns are shifting, and more extreme climate events like heavy rainstorms and record high temperatures are taking place worldwide.

This books written for academics, researchers and practitioners working in Environment field, expressed comprehensive and interdisciplinary focus on planning actions to respond to climate change.

This book provides an essential guide to researchers, it offers: various aspects of climate variation, anthropogenic factors; on the challenges and experiences in present scenario.

Simply explained, Climate Change is an important book bringing together diverse viewpoints from Industries and state agencies and regulators, for all who wish to make a difference in how to conserve and manage our Earth's climate.

Dr. Hemant Pathak

M.Sc. (Gold medalist), Ph. D.

Assistant Professor of Engineering Chemistry

Indira Gandhi Govt. Engineering college, Sagar, MP, India

Acronymns

CH_4 methane

CO_2 carbon dioxide

CRIS Climate Registry Information System

EMS Environmental Management System

GHG greenhouse gas

GIS Geographic Information System

N2O nitrous oxide

O&M operations and maintenance

Glossary

Abatement	The reduction or elimination of pollution.
Acclimatization	The physiological adaptation to climatic variations.
Acid rain	The precipitation of dilute solutions of strong mineral acids, formed by the mixing in the atmosphere of various industrial pollutants
Adaptation	Adjustment in natural or human systems to a new or changing environment
Aerosol	Particles of solid or liquid matter than can remain suspended in air from a few minutes to many months depending on the particle size and weight.
Afforestation	Planting of new forests on lands that historically have not contained forests
Air pollution	Toxic or radioactive gases or particulate matter introduced into the atmosphere, usually as a result of human activity.
Anthropogenic	Resulting from or produced by human beings.
Aquifer	A stratum of permeable rock that bears water.
Atmosphere	The gaseous envelop surrounding the Earth.
Biodiversity	The numbers and relative abundances of different genes (genetic diversity), species, and ecosystems (communities) in a particular area.
Black carbon	Operationally defined species based on measurement of light absorption and chemical reactivity and/or thermal stability; consists of soot, charcoal, and/or possible light absorbing refractory organic matter.
Bioenergy	The conversion of biomass into useful forms of energy such as heat, electricity and liquid fuels.

Biogas	Gas produced by the biological process of anaerobic (without air) digestion of organic material.
Biomass	Organic, non-fossil material of biological origin constituting an exploitable energy source.
Carbon dioxide (CO_2)	Carbon dioxide is a product of fossil-fuel combustion as well as other processes. It is considered a greenhouse gas as it traps heat (infrared energy) radiated by the Earth into the atmosphere and thereby contributes to the potential for global warming.
Carbon footprint	The amount of carbon an entity of any type (e.g., person, group, vehicle, event, building, corporation) emits into the atmosphere
Carbon sink	A reservoir that absorbs or takes up released carbon from another part of the carbon cycle. The four sinks, are the atmosphere, terrestrial biosphere, oceans, and sediments (including fossil fuels)
Climate	Climate in a narrow sense is usually defined as the "average weather"
Climate change	A regional change in temperature and weather patterns. Current science indicates a discernible link between climate change over the last century and human activity, specifically the burning of fossil fuels.
Climate model	A numerical representation of the climate system based on the physical, chemical, and biological properties of its components, their interactions and feedback processes, and accounting for all or some of its known properties.
Climate prediction	A climate prediction or climate forecast is the result of an attempt to produce a most likely description or estimate of the actual evolution of the climate in the future

Climate system	The climate system is the highly complex system consisting of five major components: the atmosphere, the hydrosphere, the cryosphere, the land surface and the biosphere, and the interactions between them.
Clean Energy Technologies	Electricity and/or heat producing systems that produce negligible or minimal amounts of environmental pollution compared with conventional technologies.
Coal	A readily combustible black or brownish-black rock whose composition, including inherent moisture, consists of more than 50 percent by weight and more than 70 percent by volume of carbonaceous material. It is formed from plant remains that have been compacted, hardened, chemically altered, and metamorphosed by heat and pressure over geologic time.
Combustion	Burning. Many important pollutants, such as sulfur dioxide, nitrogen oxides, and particulates (PM-10) are combustion products, often products of the burning of fuels such as coal, oil, gas, and wood.
Contamination	The act of polluting or making impure; any indication of chemical, sediment, or biological impurities.
Dust	Solid particulate matter that can become airborne.
Ecosystem	An interactive system that includes the organisms of a natural community association together with their abiotic physical, chemical, and geochemical environment.
El Niño Southern Oscillation (ENSO)	El Niño, is a warmwater current. This oceanic event is associated with a fluctuation of theintertropical surface pressure pattern and circulation in the Indian and Pacific Oceans, called the Southern Oscillation.
Emission	Release of pollutants into the air from a source. We say sources emit pollutants. Continuous emission monitoring systems (CEMS) are machines, which some large sources are required to install, to make continuous measurements of

pollutant release.

Erosion	The process of removal and transport of soil and rock by weathering, mass wasting, and the action of streams, glaciers, waves, winds, and underground water.
Exposure	The nature and degree to which a system is exposed to significant climatic variations.
Glacier	A mass of land ice flowing downhill and constrained by the surrounding topography
Geothermal	Natural heat extracted from the earth's crust using its vertical thermal gradient, where discontinuity in the earth's crust
Global warming	increase in the average temperature of the earth's surface.
Greenhouse gases	Atmospheric gases such as carbon dioxide, methane, chlorofluorocarbons, nitrous oxide, ozone, and water vapor that slow the passage of re-radiated heat through the Earth's atmosphere.
Hydrocarbons	An international agreement adopted in December 1997 in Kyoto, Japan. The Protocol sets binding emission targets for developed countries that would reduce the emissions on average 5.2 percent below 1990 levels.
Industrialized countries	The metric prefix for one millionth of the unit that follows.
Impact assessment (Climate)	The practice of identifying and evaluating the detrimental and beneficial consequences of climate change on natural and human systems.
Kyoto Protocol	An international treaty created in 1997 in Kyoto, Japan to reduce industrial nation's global emissions of greenhouse gases.
Methane (CH4)	A gas emitted from coal seams, natural wetlands, rice paddies, enteric fermentation (gases emitted by ruminant animals), biomass burning, anaerobic decay or organic wastes in landfill sites, gas drilling and the activities of termites.

Phytoplankton	Phytoplankton are the dominant plants in the sea, and are the bast of the entire marine food web.
Rapid climate change	The non-linearity of the climate system may lead to rapid climate change, sometimes called abrupt events or even surprises.
Solar radiation	Radiation emitted by the Sun. It is also referred to as shortwave radiation. Solar radiation has a distinctive range of wavelengths (spectrum) determined by the temperature of the Sun.
Zooplankton	The animal forms of plankton. They consume phytoplankton or other zooplankton.

1. Introduction

The earth's climate system consists of five components viz. atmosphere, hydrosphere, cryosphere, lithosphere and biosphere. While Climate change is a long-term shift in the statistics of the weather patterns. This refers to major changes in temperature, rainfall, snow, or wind patterns lasting for decades or longer. Both human-made and natural factors contribute to climate change.

Global climate is currently changing. Many changes have been observed in over the past century and the beginnings of the 21st have been the warmest period in the entire global temperature. Scientists actively work to understand past and future climate by using observations and theoretical models.

The Fourth Assessment Report of the IPCC concludes, that most of the observed increase in the globally averaged temperature since the mid-20th century is very likely due to the observed increase in anthropogenic greenhouse gas concentrations.

The large climate change has happened over a very short timeframe, and mounting evidence indicates it has the potential to radically alter marine ecosystems, as well as the health of coral reefs, shellfish, and fisheries.

Climate change is caused by factors such as certain human activities like Global warming.

- **Human causes** include burning fossil fuels, cutting down forests, and developing land for farms, cities, and roads. These activities all release greenhouse gases into the atmosphere.

- **Natural causes** include changes in the Earth's orbit, the circulation of the ocean and the atmosphere, biotic processes, variations in solar radiation received by Earth, plate tectonics, and volcanic eruptions.

A climate record has been assembled, and continues to be built up, based on geological evidence from borehole temperature profiles, cores removed from deep accumulations of ice, floral and faunal records, glacial processes, stable-isotope and other analyses of sediment layers, and records of past sea levels.

More recent data are provided by the instrumental record based on the physical sciences, are often used in theoretical approaches to match past climate data, make future projections, and link causes and effects in climate change.

2. Human Causes

Naturally occurring gases, Like CO_2 and water vapor, trap heat in the atmosphere causing a greenhouse effect. Burning of fossil fuels, like oil, coal, and natural gas followed by aerosols (particulate matter in the atmosphere) and the CO_2 released by cement manufacture is adding CO_2 to the atmosphere.

Increase in CO_2 levels due to emissions from fossil fuel combustion, Other factors, including land use, ozone depletion, animal agriculture and deforestation, are also of concern in the roles they play both separately and in conjunction with other factors in affecting climate, microclimate, and measures of climate variables.

The current CO_2 level is the highest in the past 650,000 years. The Earth's climate depends on the functioning of a natural "greenhouse effect." This effect is the result of heat-trapping gases like water vapor, carbon dioxide, ozone, methane, and nitrous oxide, which absorb heat radiated from the Earth's surface and lower atmosphere and then radiate much of the energy back toward the surface.

Without this natural greenhouse effect, the average surface temperature of the Earth would be about 60°F colder. However, human activities have been releasing additional heat-trapping gases, intensifying the natural greenhouse effect, thereby changing the Earth's climate. Climate is influenced by a variety of factors, both human-induced and natural. The increase in the carbon dioxide concentration has been the principal factor causing warming over the past 50 years. Its concentration has been building up in the Earth's atmosphere since the beginning of the industrial era in the mid-1700s, primarily due to the burning of fossil fuels (coal, oil, and natural gas) and the clearing of forests. Human activities have also increased the emissions of other greenhouse gases, such as methane, nitrous oxide, and halocarbons. Levels of carbon dioxide in the atmosphere stayed in a narrow range over the last million years. In the last hundred they have risen from 280 ppm to 400 ppm. Changes in CO_2 concentrations closely match human emissions from burning coal, oil, and gas.

Some of these activities partially offset the warming caused by greenhouse gases, while others increase the warming. One such influence on climate is caused by aerosols; the burning of coal produces emissions of sulfur-containing compounds form sulfate aerosol, which reflect some of the incoming sunlight away from the Earth, causing a cooling influence at the surface. Sulfate aerosols also tend to make clouds more efficient at reflecting sunlight, aerosols can either mask or increase the warming caused by increased levels of greenhouse gases.

Human activities have also changed the land surface in ways that alter how much heat is reflected or absorbed by the surface. Such changes include the cutting and burning of forests, the replacement of other areas of natural vegetation with agriculture and cities, and large-scale irrigation. These transformations of the land surface can cause warming or cooling.

3. Natural causes

Sun and volcanic eruptions is two important natural factors that influence climate. Global temperatures have risen sharply. Sun's energy output has followed its historical 11-year cycle of small ups and downs, but with no net increase. Solar output

The Sun is the predominant source of energy input to the Earth. Both long- and short-term variations in solar intensity are known to affect global climate.

El Niño phenomenon has important influences on global climate. Many other modes of variability have been identified by climate scientists and their effects on climate occur at the same time as the effects of human activities, the Sun, and volcanoes.

The ocean is a fundamental part of the climate system, some changes in it occurring at longer timescales than in the atmosphere, massing hundreds of times more and having very high thermal inertia. On longer time-scales, alterations to ocean processes such as thermohaline circulation play a key role in redistributing heat by carrying out a very slow and extremely deep movement of water and the long-term redistribution of heat in the world's oceans.

4. Climate forcing mechanisms

The Factors shapes the climate known as climate forcing factors. Forcing mechanisms are of 2 types-

I. Internal
II. External

Internal forcing mechanisms are natural processes within the climate system itself like the thermohaline circulation while External forcing mechanisms can be either natural (changes in solar output) or anthropogenic (Emissions of greenhouse gases).

These include some processes like variations in solar radiation, variations in the Earth's orbit, variations in the reflectivity of the continents and oceans, mountain-building and continental drift and changes in GHG concentrations.

The oceans and ice caps, respond more slowly in reaction to climate forcings, while others respond more quickly. There are also key threshold factors which when exceeded can produce rapid change.

Whether the initial forcing mechanism is internal or external, the response of the climate system might be fast viz. a sudden cooling due to airborne volcanic ash reflecting sunlight, slow like thermal expansion of warming ocean water, or a combination viz. quick loss of albedo in the arctic ocean as sea ice melts, followed by more gradual thermal expansion of the water.

Therefore, the climate system can respond abruptly, but the full response to forcing mechanisms might not be fully developed for centuries or even longer.

I. **Internal forcing mechanisms**

Natural changes in the climate system result in internal climate variability viz. type and distribution of species, and changes in ocean currents.

Life affects climate through its role in the carbon and water cycles and through such mechanisms as albedo, evapotranspiration, cloud formation, and weathering.

II. **External forcing mechanisms**

Volcanoes are also part of the extended carbon cycle. Over very long (geological) time periods, they release carbon dioxide from the Earth's crust and mantle, counteracting the uptake by sedimentary rocks and other geological carbon dioxide sinks.

Volcanoes are indistinct part of the lithosphere and climate system and volcanic emissions are at a much lower level than the effects of anthropogenic activities, which generate many times amount of carbon dioxide emitted by volcanoes.

Motion of tectonic plates reconfigures global land and ocean areas and generates topography. This can affect both global and local patterns of climate and atmosphere-ocean circulation.

The locations of the seas are important in controlling the transfer of heat and moisture across the globe, and therefore, in determining global climate.

Glaciers are the most important indicators of climate change. Their size is determined by a mass balance between snow input and melt output. As temperatures warm, glaciers retreat unless snow precipitation increases to make up for the additional melt. Glaciers grow and shrink due both to natural variability and external forcings.

5. Effects of Climate Change

There are numerous potential impacts on climate change includes the effects on sea level, drought, local weather, and hurricanes. Emissions of GHG thickening the blanket of heat-trapping gases in Earth's atmosphere, causing climate change.

Many of these projected impacts pose serious risks to human societies and things people care about, including water resources, coastlines, infrastructure, human health, food security, and land and ocean ecosystems.

Global warming is the cause of ocean warming. More than 90% of the excess energy coming to the Earth from greenhouse gases goes into the ocean waters.

Some important effects described as follows-

a. Arctic sea ice loss

Arctic sea ice is decline, ice melts away and reforms annually. It covers millions of square miles in the north and south pole, varying with the seasons. Arctic sea ice is now declining both in extent and thickness at a rate of 13.3 percent per decade.

b. Effect on Vegetation and animals

Changes in climate results in increased precipitation and warmth consequently improved plant growth and the increased amount of CO_2 extends distribution and coverage of vegetation. A gradual increase in warmth in a region will lead to earlier flowering and fruiting times, driving a change in the timing of life cycles of dependent organisms.

As the climate has changed, many species have shifted their range toward the poles and to higher altitudes as they try to stay in areas with the same ambient temperatures. The timing of different seasonal activities is also changing. Several plant species are blooming earlier in spring, and some birds, mammals, fish, and insects are migrating earlier, while other species are altering their seasonal breeding patterns.

c. Ocean acidification

CO_2 reacts in seawater to form carbonic acid, the acidification of the world's oceans is another certain outcome of elevated CO_2 concentrations in the atmosphere. seawater is becoming less alkaline through a process generally referred to as ocean acidification. The pH of seawater has decreased and projected to drop much more dramatically by the end of the 21st century if carbon dioxide concentrations continue to increase. Such ocean acidification is essentially irreversible over a time scale of centuries.

Ocean acidification affects the process of calcification by which living things create shells and skeletons, with substantial negative consequences for coral reefs, mollusks, and some plankton species important to ocean food chains.

d. Sea level is rising

Sea level rose by roughly 8 inches over the past century. Satellite data available over the past 15 years show sea level rising at a rate roughly double the rate observed over the past century. In 1996, the IPCC Second Assessment Report cautiously concluded that "the balance of evidence suggests a discernible human influence on global climate. Since then, a number of national and international assessments have come to much stronger conclusions about the reality of human effects on climate.

Altimeter measurements with accurately determined satellite orbits have provided an improved measurement of global sea level change. To measure sea levels prior to instrumental measurements, scientists have dated coral reefs that grow near the surface of the ocean, coastal sediments, marine terraces and nearshore archaeological remains. The predominant dating methods used are uranium series and radiocarbon, with cosmogenic radionuclides being sometimes used to date terraces that have experienced relative sea level fall.

Some of Earth's most densely populated regions lie at low elevation, making rising sea level a cause for concern. Sea-level rise is projected to continue for centuries in response to human caused increases in greenhouse gases, with an estimated 0.5-1.0 meter (20-39 inches) of mean sea-level rise by 2100. However, there is evidence that sea-level rise could be greater than expected due to melting of sea ice. Recent studies have shown more rapid than expected melting from glaciers and ice sheets. Observed sea-level rise has been near the top of the range of projections that were made in 1990

e. Warming of the Earth's surface

Global average surface air temperature has increased. The estimated change in the average temperature of Earth's surface is based on measurements from thousands of weather stations, ships, and buoys around the world, as well as from satellites. Scientific assessments find that most of the warming of the Earth's surface over the past 50 years has been caused by human activities average temperature and sea level have increased, and precipitation patterns have changed.

In many areas of the globe, snow cover is expected to diminish, with snowpack building later in the cold season and melting earlier in the spring.

According to one sensitivity analysis, each 1°C (1.8°F) of local warming may lead to an average 20% reduction in local snowpack in the western United States. Snowpack has important implications for drinking water supply and hydropower production.

f. Changing in precipitation patterns

Warmer air holds more water vapor, which has led to a measurable increase in the intensity of precipitation. Changes have been observed in the amount, intensity, frequency, and type of precipitation.

precipitation is not distributed evenly over the globe. Its average distribution is governed primarily by atmospheric circulation patterns, the availability of moisture, and surface terrain effects. This are influenced by temperature. Human-caused changes in temperature are expected to alter precipitation patterns. Observations show that such shifts are occurring.

g. Droughts and flooding

There have been increases in the occurrences of both droughts and floods. As the world warms, northern regions and mountainous areas are experiencing more precipitation falling as rain rather than snow.

Increases in heavy precipitation events have occurred, even in places where total rain amounts have decreased. These changes are associated with the fact that warmer air holds more water vapor evaporating from the world's oceans and land surface. This increase in atmospheric water vapor has been observed from satellites, and is primarily due to human influences.

h. Forest fire and Changing in climate

Rising temperatures and increased evaporation are expected to increase the risk of fire in many regions of the World. Length of forest fire seasons worldwide increased by 18.7 percent due to more rain-free days and hotter temperatures. Four independent environmental factors influenced from climate change increased the possibility of wildfires-

1. hotter temperatures
2. decreased relative humidity
3. more rain-free days
4. higher wind speeds.

On all the forested continents, except Australia, the fire seasons are getting longer.

i. Climate's effect on health

The climate's effect on health is generally less pronounced in wealthier countries like the United States, where so many people are protected from the elements in their homes. But climate change is affecting health in developed countries, too.

A White House report listed deepening risks. Asthma will worsen, heat-related deaths will rise, and the number and travelling range of insects carrying diseases once confined to the tropics will increase.

j. Agriculture and food production

The stress of climate change on farming may threaten global food security. Although an increase in the amount of CO_2 in the atmosphere favors the growth of many plants, it does not necessarily translate into more food. Crops tend to grow more quickly in higher temperatures, leading to shorter growing periods and less time to produce grains. In addition, a changing climate will bring other hazards, including greater water stress and the risk of higher temperature peaks that can quickly damage crops.

Agricultural impacts will vary across regions and by crop. Moderate warming and associated increases in CO_2 and changes in precipitation are expected to benefit crop and pasture lands in middle to high latitudes but decrease yield in seasonally dry and low-latitude areas. In California, where half the nation's fruit and vegetable crops are grown, climate change is projected to decrease yields of almonds, walnuts, avocados, and table grapes by up to 40 percent by 2050.

For each degree of warming, yields of corn in the United States and Africa, and wheat in India, drop by 5-15%

• Crop pests, weeds, and disease shift in geographic range and frequency

• If 5°C of global warming were to be reached, most regions of the world would experience yield losses, and global grain prices would potentially double.

6. Climate models

For several decades, scientists have used the world's most advanced computers to simulate the

Earth's climate. These models are based on a series of mathematical equations representing the physics laws that govern the behavior of the atmosphere, the oceans, the land surface, and other parts of the climate system, as well as the interactions among different parts of the system.

Scientists use climate models to project how the climate system will respond to different scenarios of future greenhouse gas concentrations.

Climate models are important tools for understanding past, present, and future climate change. Climate models are tested against observations so that scientists can see if the models correctly simulate what actually happened in the recent or distant past.

Earth simulations computational projection predicts that, in the absence of human activities, there would have been negligible warming, or even a slight cooling, over the 20th century. When greenhouse gas emissions and other activities are included in the models, resulting surface temperature changes more closely resemble the observed changes.

In climate modelling the aim is to study the physical mechanisms and feedbacks of volcanic forcing. surface temperatures to rise. The oceans are warming faster than climate models predicted

7. Greenhouse gas reductions and alternate energy and Improved technology

To control on climate change in the long term, the most important greenhouse gas to control is carbon dioxide, emitted as a result of burning fossil fuels. Amount of emissions from residential, commercial, industrial, and transportation sources.

The United States is responsible for about half of the human-produced CO_2 emissions already in the atmosphere and currently accounts for roughly 20% of global CO_2 emissions, despite having only 5% of the world's population. The U.S. percentage of total global emissions is projected to decline over the coming decades as emissions from rapidly developing nations such as China and India will continue to grow. Thus, reductions in U.S. emissions alone will not be adequate to avert climate change risks.

Expand education and incentive programs to influence consumer behavior and preferences; curtail sprawling development patterns that further our dependence on petroleum.

A report said, Improving the energy efficiency of air conditioners could save up to 100 billion tonnes of CO_2 by 2050, according to new research from the Institute for Governance and Sustainable Development (IGSD).

IGSD highlights the benefits of improving the efficiency of air conditioning units will slash future carbon emissions. The findings from the Lawrence Berkeley National Laboratory in California also call for the parallel phase out of hydroflourocarbon (HFC) refrigerants under the ongoing Montreal Protocol.

Expand the use of low- and zero-carbon energy sources, for example, switch from coal and oil to natural gas, expand the use of nuclear power and renewable energy sources such as solar, wind, geothermal, hydropower, and biomass; capture and sequester CO_2 from power plants and factories.

Capture and sequester CO_2 directly from the atmosphere, for example, manage forests and soils to enhance carbon uptake; develop mechanical methods to "scrub" CO2 directly from ambient air.

Advancing these opportunities to reduce emissions will depend to a large degree on private sector . In general, there are four major tool chests from which to select policies for driving emission reductions:

• Pricing of emissions such as by means of a carbon tax or cap-and-trade system;

• Mandates or regulations that could include direct controls on emitters (through law) or mandates such as automobile fuel economy standards, appliance efficiency standards, labeling requirements, building codes, and renewable or low-carbon portfolio standards for electricity generation;

• Public subsidies for emission-reducing choices through the tax code, appropriations, or loan guarantees; and

• Providing information and education and promoting voluntary measures to reduce emissions

8. Climate benefits

The research calculates a potential 1,200GW of electricity production could be avoided by improving global air conditioner efficiency, preventing up to 0.5 degrees of global warming alone.

The group estimates by reducing the demand on energy intensive air conditioners countries could potentially prevent thousands of power plants being constructed worldwide.

9. **Conclusion**

Industry experts have repeatedly claimed the UK and other nations will miss carbon emissions targets without sustained investment in cold energy. Climate change is about making choices in the face of risk.

Any course of action carries potential risks and costs; but doing nothing may pose the greatest risk from climate change and its impacts. However, robust scientific knowledge and analyses are a crucial foundation for informing choices.

10. References

1. Christner, B. C.; Morris, C. E.; Foreman, C. M.; Cai, R.; Sands, D. C. (2008). "Ubiquity of Biological Ice Nucleators in Snowfall". *Science* 319 (5867): 1214.doi:10.1126/science.1149757. PMID 18309078.

2. Schwartzman, David W.; Volk, Tyler (1989). "Biotic enhancement of weathering and the habitability of Earth". *Nature* 340 (6233): 457–460. doi:10.1038/340457a0.

3. Kopp, R. E.; Kirschvink, J. L.; Hilburn, I. A.; Nash, C. Z. (2005). "The Paleoproterozoic snowball Earth: A climate disaster triggered by the evolution of oxygenic photosynthesis". *Proceedings of the National Academy of Sciences* 102 (32): 11131–6. doi:10.1073/pnas.0504878102. PMC 1183582. PMID 16061801.

4. US EPA. *Glossary of climate change terms*.

5. IPCC (2007). "What are Climate Change and Climate Variability?"

6. Sagan, C.; Chyba, C (1997). "The Early Faint Sun Paradox: Organic Shielding of Ultraviolet-Labile Greenhouse Gases". *Science* 276 (5316): 1217–21.

7. "Volcanic Gases and Climate Change Overview". *http://www.usgs.gov/*

8. Human Activities Emit Way More Carbon Dioxide Than Do Volcanoes". American Geophysical Union. 14 June 2011

9. NASA Global Climate Change "Climate Change: How do we know?

ABOUT THE AUTHOR

Dr. Hemant Pathak held positions as Assistant Professor in the department of chemistry, Govt. Indira Gandhi Engineering College, Sagar, MP, India. He had extensive experience in teaching, research and administrative management.

Dr. Pathak received his Ph.D. degree in chemistry from Dr. Hari Singh Gour Central University, Sagar, India and M.Sc. Gold medalist from Jiwaji University, Gwalior. He has published 22 books and more than 50 research papers in reputed International and National journals and received several awards. He is a member of editorial boards and reviewer boards of several international journals and societies. His area of specialization includes Engineering Chemistry, Energy audits and Environmental Pollution management.

www.ingramcontent.com/pod-product-compliance
Lightning Source LLC
Chambersburg PA
CBHW081823170526
45167CB00008B/3515